Eye and Vision Research Developments

COLOR: ONTOLOGICAL STATUS AND EPISTEMIC ROLE

Eye and Vision Research Developments

Eye Cancer Research Progress
Edwin B. Bospene (Editor)
2008. 978-1-60456-045-9

Non-Age Related Macular Degeneration
Enzo B. Mercier (Author)
2008. 978-1-60456-305-4

Optic Nerve Disease Research Perspectives
Benjamin D. Lewis and Charlie James Davies (Editors)
2008. 978-1-60456-490-7

Optic Nerve Disease Research Perspectives
Benjamin D. Lewis and Charlie James Davies (Editors)
2008. 978-1-60741-938-9

New Topics in Eye Research
Lauri Korhonen and Elias Laine (Editors)
2009. 978-1-60456-510-2

Eye Infections, Blindness and Myopia
Jeffrey Higgins and Dominique Truax (Editors)
2009. 978-1-60692-630-7

Eye Research Developments: Glaucoma, Corneal Transplantation, and Bacterial Eye Infections
Alan N. Westerhouse (Editor)
2009. 978-1-60741-177-2

Retinal Degeneration: Causes, Diagnosis and Treatment
Robert B. Catlin (Editor)
2009. 978-1-60741-007-2

Retinal Degeneration: Causes, Diagnosis and Treatment
Robert B. Catlin (Editor)
2009. 978-1-60876-442-6

Binocular Vision: Development, Depth Perception and Disorders
Jacques McCoun and Lucien Reeves (Editors)
2010. 978-1-60876-547-8

Understanding Corneal Biomechanics through Experimental Assessment and Numerical Simulation
Ahmed Elsheikh (Author)
2010. 978-1-60876-694-9

Retinitis Pigmentosa: Causes, Diagnosis and Treatment
Michaël Baert and Cédric Peeters (Editors)
2010. 978-1-60876-884-4

Color: Ontological Status and Epistemic Role
Anna Storozhuk (Author)
2010. 978-1-61668-201-9

Coherent Effects in Primary Visual Perception
V.D. Svet and A.M. Khazen (Author)
2010. 978-1-61668-143-2

Conjunctivitis: Symptoms, Treatment and Prevention
Anna R. Sallinger (Editor)
2010. 978-1-61668-321-4

Conjunctivitis: Symptoms, Treatment and Prevention
Anna R. Sallinger (Editor)
2010. 978-1-61668-443-3

Coherent Effects in Primary Visual Perception
V.D. Svet and A.M. Khazen (Authors)
2010. 978-1-61668-496-9

Color: Ontological Status and Epistemic Role

Anna Storozhuk (Author)

2010. 978-1-61668-608-6

Eye and Vision Research Developments

COLOR: ONTOLOGICAL STATUS AND EPISTEMIC ROLE

ANNA STOROZHUK

Nova Science Publishers, Inc.
New York

For permission to use material from this book please contact us:
Telephone 631-231-7269; Fax 631-231-8175
Web Site: http://www.novapublishers.com

LIBRARY OF CONGRESS CATALOGING-IN-PUBLICATION DATA

Available upon Request
ISBN: 978-1-61668-201-9

Published by Nova Science Publishers, Inc. ✝ New York

CONTENTS

PREFACE

In this chapter the complex impact of color action on the organism is discussed. It is shown that the perception of color depends on the action of irritants on other sensor systems and, vice versa, the action of color may exert exciting or inhibiting influence on the perception of sounds or smells. The facts of mutual influence of the sensor systems force us to consider not only a single observation but the entire experience as a continuous flow of sensations which is formulated as the empirical monism principle. The question is raised why the opinion about the atomicity of experience or the opinion that observation is a discrete fact of fixation of the state of affairs are so widely shared. The reason seems to lie in the supposition about the existence of an isomorphism between thinking and the world, which the author calls the principle of identity of being and thinking. Since language consists of separate sentences, the world, according to the identity principle, must consist of separate facts so that the correspondence may be established. Another myth about experience, also being in accordance with the principle of the being and thinking identity, is a hypothesis about the passivity of perception and its mirror-like character. The picture theory supplied an automatic solution to some ontological and epistemological problems but encountered some other problems such as the problems concerning explanation of the objective character of perception, the space perception, the knowledge about unperceivable parts of the objects, and the ontological status of colors. We propose the solutions of these problems from the point of view of perception's activity and continuity. We discuss the mechanism of increasing realism of colored images, as well as the epistemic role of color.

INTRODUCTION

Traditionally, the problem of color realism includes the discussion of the capability of color to be a reason of visual experience of a certain type. The problem is often formulated in the form 'what does it mean to be colored?', and the possible answers to this question appeal to color perception. However, philosophers avoid discussing the very process of color action on the organism, considering perception to be a simple non-problematic source of knowledge about the world. At the same time, raising the question concerning the ways of color action allows reformulating the issue of the color and reality relations.

The main purpose of this article is to discuss precisely the mechanisms of the color perception processes and the ways of color action on the organism, during which the subject experiences the sense of reality and has a cognitive experience. Clarification of the mechanism of appearance of the subjective color experience, which is a result of influence on the organism of electromagnetic waves carrying certain energy, allows bringing to light the usually missed peculiarities of the mind-body problem. Based on the data on the ways of color action on the organism, there will be discussed the following questions: 'why does color make the perception more realistic?' and 'how does color influence the cognition process?'.

In the first part there are considered the physical properties of color discovered by Newton and interaction of various colors. It is studied then how can color, being a characteristic of electromagnetic wave and not being a property of objects, exert, nevertheless, its influence on the organism. The main attention is given to non-uniqueness of such influence. For example, a thin ray of white color, moving along the eye retina, will produce different color sensations. Besides, the action of color changes the sensitivity of other sense organs and of twilight vision. The following issues will be discussed: how do the color-sensing elements of the eye interact? How does color perception influence other sensor capabilities? What is the mechanism of transferring such influence? In this part I am using the data obtained by Russian physiologists whose work, as far as I know, has been rarely quoted. Nevertheless, the facts of mutual influence of various sensations have been known for a long time and are still being investigated by physiologists. At the same time, philosophers have been paying almost no attention to them. In the second part, we try to find the reason of philosophical inattentiveness to the adaptive changes of the sensor apparatus sensitivity. This consideration takes us to the Antique philosophy, in which there originated an orientation to identify the thought and being and then study the conceptions about the world in order to obtain knowledge. I call this orientation the principle of the being and thinking identity and show that it is the reason of so longstanding domination of the passive mirror paradigm of perception with consideration of experience as a separated isolated reaction of the organism to a singled-out irritant. These two suppositions about experience allow providing an isomorphism of cognition and existence; however, they have led philosophy into a dead-end. The views of philosophers on perception are full of myths and color perception is a good example to demonstrate irrelevancy of our conceptions about the process of acquiring experience.

I contrast the conception that our cognition is a result of separate atomic contacts with the world with *the principle of empirical holism*, in which there is postulated continuity of sensations and their complex influence on the organism. Then, from the point of view of empirical holism, there are given answers to the critical arguments against the logical empiricism and there are suggested new ways of resolving the difficulties of the passive-mirror conception: how does the sum of sensations form the bodies? How is the perception of space possible? How does color, not being an object, exert influence on the organism?

In the third part, it is shown that the influence of color on the organism can be explained only by taking into account continuity of sensations and their complex character. There is considered, from the point of view of empirical holism, the question of how does color give a greater realism to the image. Several possible mechanisms of such action are indicated, from an additional excitation of nervous system produced by colors to amplification of the organism reactions through influencing the emotional sphere. Finally, there are discussed the epistemological consequences of the perception continuity, which are illustrated by concrete examples of color influence on the organism.

Chapter 1

1. The Physical Properties of Color and Its Influence on the Organism

1.1. Newton on the Properties of Light and Color

To get an idea about the physical nature of color, let us turn to the history of optics. Isaac Newton was the first to explain the nature of color. He did it while solving a technical problem of improving telescopes. Observation through lenses was accompanied by chromatic aberration: the edges of the objects visible in the telescope were surrounded by a colored fringe degrading the picture. To find out the reason of its appearance and of the fuzziness introduced into the observation of the objects through glasses, Newton studied the mechanism of appearance of colors in the experiments which lasted more than 10 years. There existed three explanations of the nature of color at that time. The first one was due to Aristotle and connected the effect of color appearance with the boundaries of objects; Hooke presumed that color is caused by various modes of oscillation of the ether medium; whereas other philosophers believed that color is born out of shadow and light mixing. According to Newton, all these viewpoints had the same mistake of assuming that colors are not inherent in light from the beginning but acquired in reflection and refraction and generated by theses processes themselves. In particular, concerning the observation of prism's decomposition of white light into its spectra it was presumed that the colors are an effect introduced by the prism itself. However, Newton could not accept these ideas, since color effects were manifested also without bodies: in rainbow, prisms, glasses, and liquids. He demonstrated that colors are contained in white light itself, they are inherent in it, not just appearing in reflections and refractions. To verify this, Newton decomposed first a light ray with the help of a prism and then recomposed the rays using a lens and a second prism. In both cases there was obtained the same white light as before the decomposition, which means that the prismatic colors do not disappear while forming whiteness but only mix together. Decomposition of white light into its spectra is possible due to the fact that the rays with different wavelengths have different refraction coefficients. Therefore, the bodies lighted by the composite light, seem to be fuzzy and colored; whereas the bodies lighted by homogeneous light seem distinct and are colored the same way as they are if you look at them simply by naked eyes.

> DEFIN. VII 'The Light whose Rays are all alike Refrangible, I call Simple, Homogeneal and Similar; and that whose Rays are some more Refrangible than others, I call Compound, Heterogeneal and Dissimilar.
>
> DEFIN. VIII The Colours of Homogeneal Lights, I call Primary, Homogeneal and Simple, and those of Heterogeneal Lights, Heterogeneal and Compound' [1].

If we make the colors, obtained by decomposition, pass through a second prism, then the colored ray does not decompose further; i.e. this color is 'pure' or 'simple' because it contains the rays with one and the same wavelength. It also gave witness to the fact that the colors are not an effect of refraction in the prism but a property of the rays. Newton came to a conclusion that the prism serves just to separate the rays already present in the original light. At the same time, Newton understood that color arises as a result of interaction between the light radiation and our sensuous ability; the rays do not have colorings on their own.

Newton, following Descartes, supposed light to be corpuscles; but explained differently the reason of color sensation. Newton considered the speed of the particles to be the cause of our experiences of color, while Descartes thought it had to do with their rate of rotation [2].

Later, when instead of Newton's corpuscular optics, Ch. Huygens, T. Young, and O. Fresnel developed wave optics, Descartes' explanation of the color sensation became more accepted and colors started to be explained by the differences in wavelength.

> The sensation of different colors depends on the different frequency of vibrations excited by light in the retina... Cause a sensation of light by beating and dashing against the bottom of the eye, something after the manner that vibrations in the air cause a sensation of sound by beating against the organs of hearing. Now, the most free and natural application of this hypothesis to the solution of phenomena I take to be this that the agitated parts of bodies, according to their several sizes, figures, and motions, do excite vibrations in the ether of various depths or bignesses, which, being promiscuously propagated through that medium to our eyes, effect in us a sensation of light of a white color; but if by any means those of unequal bignesses be separated from one another, the largest beget a sensation of a red color; the least, or shortest, of a deep violet, and the intermediate ones of intermediate colors [3].

The explanation, given to the physical nature of color as wavelength, is basic nowadays. However, the subjective experience of color does not coincide with its objective characteristics as an electromagnetic wave of given energy. The perception of color may depend both on physical conditions (brightness, intensity, wavelength) and psychological readiness to perceive it. Although the color that we see depends on the characteristics of the rays coming to eyes, the connection between the color impression and physical composition of the stimulant is not simple. The rays, being different by composition, may cause absolutely identical observations of color. For example, white color can be obtained by mixing various pairs of complementary colors, the monochromatic yellow cannot be told apart from the mixture of red and green rays.

Besides, one and the same stimulant may give different sensations of color. For example, if a narrow white colored ray gets to various segments of the eye's retina, then, depending on the position of the ray, its color will change from red in the central part of the eye and through orange and yellow to blue and violet spot in the ray's center on the periphery of the eye.

Non-uniqueness of the correspondence between the light stimulant and reaction to it does not allow giving an unambiguous definition of color and requires taking into account a wide range of phenomena such as general picture of the excitation of the color-sensitive apparatuses and their interaction.

1.2. Interaction of the Color-Sensing Elements of the Eye

A general picture of activation of the color-sensing apparatuses is given by the tristimulus theory, which presumes the presence of red-green- and blue-sensing cone cells in the eye retina. Color perception is connected with the fact that the absorbed photons stimulate decomposition of various pigments reacting in the cone cells due to incoming light. The perception of other colors is a result of their interaction; however, a number of experiments bear witness to the phenomena that the perceptions of monochromatic light and of the corresponding mixture of colors are not identical.

For example, according to the tristimulus theory, the yellow color perception depends on the excitation of the green- and red-sensing apparatuses of our color vision and is not independent. However, a number of observations show relative independence of yellow color from the excitation by green and red. For instance, after irritation of the eye by monochromatic yellow, its sensitivity to perception of blue color increases; whereas, after irritation by a mixture of red and green, it does not [4]. Another effect is connected with the fact that the boundaries of the peripheral vision's domain for yellow color are wider than for green. Hartridge (1947) demonstrated that the presence of a yellow spot in the vision field increases the sensitivity of color resolution for the irritants of small angular size. The presence of red and green spots yielded no change in the visibility threshold of the control color spot.

Non-uniqueness of the reaction manifests also in non-equal time of reaction to various color irritants. Some colors excite faster the reacting apparatus of the retina's rod cells, others affect the slowly reacting cone cells; the difference of the sensation time for rod and cone cells results in the difference of perception of brightness. Therefore, for two colors of the same brightness but different hue the rate of the sensation appearance is different [5].

Piéron (1945) discovered that the rate of the blue color perception is minimal, whereas the perception of red and green colors has the maximal rate. The rate of sensation appearance was increasing for the red and green irritant differently and varied depending on adaptation since green color wearies the eye less. Different rate of the color sensations arising and developing leads to some interesting phenomena. For example, a disc with black and white sectors, which has been put into rotation, will be perceived as reddish, while the color distribution will depend on the rotation velocity. This is the result of the fact that the red rays, being components of the white color reflected from the white sectors, are perceived faster than the blue ones. A time difference arises also in the process of attenuation of after-images of different colors, the red excitation attenuating the fastest of all. If we compare the yellow and white irritation in the darkness conditions, then it will turn out that the subsequent image of the yellow irritation fades away faster, which serves as an argument for the introduction of yellow lights in cars [6]. Finally, the inhibiting influence of the cone-cell apparatus on the rod-cell one is widely known.

The lighting of the central pit of the retina by red, blue and green colors results in diminishing excitability of the peripheral vision to a significantly greater degree than the lighting by the mixed white light [7].

1.3. The Color Perception and the Mutual Interaction of Various Systems of the Organism

Colors exhibit a complex influence on the organism, which is not limited to the action on the visual apparatus but they excite or inhibit the abilities of perceiving the signals from other sense organs and also the thinking processes. There is also a backward influence when the irritations of other afferent systems may lead to the change in the color perception.

The question about the afferent systems interaction has been studied for a long time. For example, the first mentioning of this effect is due to Urbantschisch (1888), who asked the test persons to look at colored spots from such distance that the colors could not be identified. With the tuning-fork sound, the previously undistinguishable colors would become clearly visible. V. V. Lazarev (1904) demonstrated the dependence of the perceived sound volume on the illumination. He would, by turns, illuminate and darken a screen in front of the audience; and those present would hear distinctly that during the darkening of the screen the pitch-fork sound became quieter than on the light. Although the early researchers had established the facts of mutual influence of various afferent systems, they neither discovered regularities in the reactions nor indicated possible mechanisms of mutual interrelation of various afferent systems. The question concerning interrelation of the stimuli of various modalities was considered more specifically by the academician L.F. Orbeli (1935), whose collaborates had conducted many experiments trying to uncover such influence, in particular, regarding the action of color on the organism.

Depending on the color of the irritant acting on the eye, there occur some changes in the condition of both the eyes and the organism as a whole. Consider specific examples. Zaretskaya (1941) discovered that lighting one eye by green results in decreasing the intraocular pressure in the other eye, which amount to 80 percent from the initial level after 20 minutes after the exposition's beginning. Lighting the eye by red color of the same brightness increases the intraocular pressure in the eye being studied. She discovered also that under the adaptation to green the blind spot decreases; under the adaptation to red, on the contrary, grows.

The general physiological condition noticeably influences color vision. Hypoxia significantly changes the thresholds of color vision, almost without any influence on the twilight one. The oxygen shortage, corresponding to the height of 5000 meters, leads to increasing sensitivity to red and yellow, diminishing the sensitivity to blue and green [8].

The experiments by O. A. Dobryakova (1948) showed that general stimulation of the central nervous system, for example, an intense intellectual work, increases the sensitivity not only to the visual irritants but also to the acoustic, osmatic, and gustative ones.

The color-perceiving apparatus is connected also with the auditory one. For example, L. A. Schwartz (1949) demonstrated that adaptation of eyes to green color increases the auditory sensitivity; whereas, to red color, diminishes.

Kravkov (1934) investigated the changes of sensitivity of color vision in its dependence on the character of the sound and showed that, with sounds, the sensitivity to green and blue increases, whereas to red and orange decreases in proportion to sound's volume. In his laboratory, there were discovered analogous effects for the boundaries of color field: the boundaries of green and blue fields are getting wider with sounds, whereas of orange one, narrower.

Later, Kravkov observed that "in the conditions of loud noise produced by the aircraft engine, light sensitivity of twilight vision dropped down to 20 percent of its level in the quiet conditions before the start of auditory irritation" [9]. Semenovskaya demonstrated that, after the cessation of sound, light sensitivity of the peripheral vision increases above the norm. The same effect is produced by an easy muscular work. Besides, Semenovskaya (1933) showed that lighting the eye by red light increases the sensitivity of twilight vision, whereas loud noise decreases it.

There were investigated also the influences of osmatic and gustative irritations on vision. For example, L. I. Seletskaya (1941) established that under the influence of rosemary smell the boundaries of the vision field for green are getting wider, whereas for red, conversely, they get narrower; and under the influence of indole smell there was observed a reverse narrowing of the vision field boundaries for green and widening, for red.

The research by Dobryakova (1944) of the gustative sensitivity showed that the sugar taste increases the color sensitivity to the region of blue-green rays and decreases it to the orange-red ones. Some changes in color vision arise also under the presence of stimuli from the proprioceptive apparatus. Schwartz (1946) established the influence of the proprioceptive irritants: with the head thrown back, the sensitivity to green color of 520 mm perceivably decreases, reaching after 90 minutes the 25 percent of the initial value; whereas, the sensitivity to orange-red somewhat increases. The muscular working capacity increases under green illumination and drops under red one.

The above mentioned studies show that a single perception influences the functioning of many other system of the organism. As applied to color, numerous experiments showed that under the influence of non-direct irritants (sounds, smells, chemical substances) there is observed a significant change of color perception of green-blue and orange-red rays, which are situated nearer to the center of the perceivable range; whereas, for the boundary rays of the spectrum – red and violet – it remains unchanged.

> For example, we saw that the auditory irritants increase the sensitivity of the dark-adapted eye in relation to orange-red rays; illumination of eyes make the sounds we hear louder; the tactile irritation of the hand by a weak electric current diminishes the auditory sensitivity; the hearing thresholds may change under the influence of osmatic sensations; illumination of eyes is favorable for the tactile sensitivity [10].

Single perceptions change the sensitivity of all sense organs, which indicates that the organism functions as a united, integral system. Under the irritation of one of sense organs, changes of the sensitivity in other sense organs sometimes take place after some time and develop during several minutes; and the effect from the irritant action is accompanied by a long aftereffect. At times, a secondary sensor irritant would give start to the sensitivity changing lasting more than half-hour and reaching maximum at the 20th minute and then the process would recede.

Schwartz (1948) proposed a hypothesis that the character of the irritant's influence is connected to the emotional condition, which accompanies the action of the irritant. To check her idea, she studied the influence of pleasant and disagreeable accords (consonance and dissonance) on the color perception and found that not only the action of sound but also the recollection about it produce very similar effects (see the table below reproduced from Schwartz, p. 317).

Influence of the recollection about the irritant on the subsequent sensitivity of color vision

Secondary irritant	Color	Time of the aftereffect, minutes			
		1	10	20	30
Consonance (major third)	Red	130	134	134	113
	Yellow	131	157	158	135
	Green	64	60	61	68
	Blue	63	62	65	75
Dissonance (minor second)	Red	70	66	63	79
	Yellow	71	65	62	73
	Green	153	186	199	173
	Blue	124	187	150	129
Recollection of the consonance	Red	126	183	139	120
	Green	69	62	60	69
Recollection of the dissonance	Red	64	68	65	75
	Green	140	159	179	123

(The background is taken as 100 percent, the irritant action was during 2 min.).

In order to establish whether it is the emotional mindset that plays the role, not the action of the auditory irritant, she conducted another series of experiments, where the test persons were asked to recollect a joyful or sad event which played a significant role in their personal life. The results also confirmed the connection between the pleasant feelings and increasing sensitivity to red color, as well as between the unpleasant ones and increasing sensitivity to green color.

Summing up the discussion on mutual influences of various systems of the organism, we should indicate common features in the processes of the color sensitivity reaction to secondary irritants. The common in the action of various irritants on the organism is the character of changes taking place in the vegetative nervous system. Such irritants as sound, the smell of bergamot oil, camphor, rosemary, a sweet taste, slight thermal irritation activate the sympathetic branch of the nervous system; whereas, the indole smell and the thrown-back position of the head stimulate the parasympathetic branch.

The vegetative system has a central character and the vegetative shifts are one of the ways of influencing color vision. Let us consider in more detail the possible mechanisms of mutual interrelation of sense organs.

1.4. The Mechanisms of Mutual Influence of Sense Organs

The question about the mechanism of influencing was studied in detail by S. V. Kravkov; and in the sequel I will keep to his classification of the mutual influence factors.

Ephaptic Connections

Excitation of a certain nervous fiber is able to influence the condition of the adjacent nerves. Such way is called the influence through touch and was confirmed by juxtaposition of the encephalograms taken from the neighboring nerves at the moment of excitation of one of them and in the period of rest. Besides the direct excitation of the close-by nerves, the excited fiber may influence the irritation thresholds; i.e., the excitability of the nearby nerves [11].

Irradiation Effect. The Rule of Leveling and Exaggeration

The influence of secondary irritant may manifest as an increase of sensitivity to some types of signals or decrease of sensitivity to some others. For example, sounds increase the vision acuity in distinguishing dark objects in front of a light background and diminish the ability to distinguish light object on a dark background. To explain such facts it was necessary to acknowledge non-uniformity of distribution of the additional excitation: it 'comes in a larger degree where there is already sufficient excitation and in a smaller degree to the less excited segments of our visual apparatus' [12]. This regularity is close to the dominant principle formulated by Ukhtomskii: 'the leading locus of excitation … accumulates in itself the excitation from the distant' [13].

Connections between Centers

There are a large number of associating fibers between various segments of the cerebral cortex which may both stimulate and inhibit activity of the centers connected with them. For example, color vision exerts the inhibiting influence on the peripheral twilight vision by lowering sensitivity of the rod cells. In the experiments, where the test person was suggested, under hypnosis, the presence of irritations, there was demonstrated that the change of sensitivity depends in a greater degree on the processes in the nervous centers, not in the nerves.

The Role of the Vegetative Nervous System

The vegetative nervous system influences the functional changing of all the excitable tissues including the sense organs. L.A. Orbeli demonstrated that the vegetative nervous system plays an important role in regulating functional systems of the organism. Since all the sense nerves pass through hypothalamus where the vegetative centers are located, irritation of any receptor may cause, besides a specific sensation, some changes throughout the entire organism: the heart rate increase, the change of blood pressure or breathing frequency and so on. The changes of the vegetative system have an extensive character, touching the entire organism or its large segments; therefore, the vegetative system is one of the most important ways through which some sense organs can influence others. For example, the influence of the green- and blue-sensing apparatuses of vision is connected with the sympathetic nervous system; whereas, the orange-red influence the parasympathetic one. Correspondingly, any influences on the corresponding system change the color sensitivity; for example, throwing back of the head increases sensitivity to red color and decreases, to green.

Sensor Conditioned Reflexes

The associative connections increase the sensations being expected or even generate them. For instance, a sight of an unlit cigarette may create illusion of the tobacco smell.

The Changing of Physiological Readiness of the Organism to Perception

Under the influence of irritation of one sense organ, there may change a condition which determines the readiness of perception of another sense organ. For example, sharp sound causes pupils narrowing which results in changing the conditions of light getting to the light-sensitive elements of the retina. The changes of the sensitivity thresholds of a sense organ under the influence of secondary irritations can be put into the same class.

Distinguishing the above mechanisms is a theoretical abstraction. In reality there always take place all the mentioned mutual influences, so that it is impossible to draw exact borders. 'In such cases, there usually takes place restructuring of the adaptation reactions, which may create unfavorable conditions for those irritants which are not in the focus of attention at the given moment' [14].

It should be indicated that each second our body receives hundreds of irritations coming both from interoception, muscular-articular sense, vestibular system and so on, and from exteroceptors. In particular, there was experimentally established the influence of unheard sounds and invisible light irritants. Taking into account that 'our sensations may ... change significantly depending on the action of secondary irritants to the subject, so weak on their own that they cannot produce any sensations' [15], perception should be considered integrally, as a systemic process causing the global changes in the organism; not as an isolated separate act. All that was said above on the character of color action on the organism and mutual influence of various sensor systems on each other requires revising our conceptions concerning perception as a simple fixation of experience's data. Activity of perception has lately become a widely discussed topic; however, in what follows I will discuss the problems connected with another myth about perception: thinking about experience as an isolated atomic fact. Experience should be considered not as isolated acts of vision and hearing, but as a continuous process of mutually interrelated sensations. From my point of view, all misconceptions related to experience and, therefore, the cognition processes in general, have a common source which will be discussed next.

Chapter 2

2. The Source of the Myths about Experience: The Principle of the Being and Thinking Identity

The topic of this section concerns discussion of the general foundations of our epistemological preferences; thus, it is somewhat abstract and, nolens volens, will take us away from our discussion of color, though they provide a nice example of inadequacy of our notions about experience. Probably, the reader has already noticed that the physiological research on color action on the organism, whose results were mentioned above, is rather old. In fact, the question of integration of sense organs was raised and experimentally studied already in the end of the 19th century and, by the middle of the 20th century, significant factual material had been accumulated. However, the physiologists' results had not been grasped and taken into account by philosophers, who are still considering experience as photography, as a simple objective source of knowledge about the world. The goal of this section is to uncover the reasons of such inertia of philosophical thinking.

In my opinion, the main source of the philosophical myths about experience is an orientation, formed in antiquity, toward studying thinking as the main source of knowledge about the world. This orientation was first formulated by Parmenides, according to whom, to think and to be is one and the same. I will refer to this orientation as *the principle of the being and thinking identity*. This principle was used implicitly, is still being applied, and has been of bad service: due to it, philosophy deviated from studying nature, having substituted it by studying the thoughts about nature. The goal of a brief review below is to familiarize the reader with the history of this principle.

1.1. The History of the Principle of the Being and Thinking Identity

Parmenides

The goal of Parmenides was to solve the problems resulted from Heraclitus' dynamics of nature. One of those problems was the problem of existence, because in order to exist the becoming should each moment arise from non-being and disappear into non-being. The second problem was the cognitive problem, because if we are to cognize the constantly changing world, then knowledge is impossible. To resolve this issues, Parmenides distinguishes a way of truth and a way of opinion; the former being a way of thinking, the second, of sensual cognition.

Both the possibility of cognition of the true existence and the existence of this true existence is derived from thinking. 'The one [way] that it is and that it is not possible not to be' [16]. It is because any really thinking object cannot not to exist, at least in thinking or as a thought; therefore, what is thought exists. Then, Parmenides makes a conclusion:

> For thought and being are the same. (B 3)
> *Thinking and the thought that it is are the same; for you will not find thought apart from what is, in relation to which it is uttered.* (B 8.) [17]

The identity of thinking and being allowed Parmenides moving away from dynamism and fluidity of things and substantiating the existence of the world and its cognizability; however, this principle put an end to natural philosophy. Since that time, studying thinking and studying language had become the main task of philosophy.

Plato

Plato considered 'Ideas' or 'Forms' to be the source of existence and cognition, while ascribing the properties of fluidity and non-being to matter which in conjunction with Ideas becomes the source of ordinary things. Since knowledge was connected with Ideas, Plato assigned a great importance to searching the definitions of the thing. 'To know that ... F –, one must be able to say what the F, or Fness, is' [18]. The definition of knowledge according to Plato implicitly presupposes the identity of being and thinking, since expressibility of things in language became the criterion of knowing things.

The supposition 'to know is to be able to say' forms a research orientation, having been perfected by his pupil Aristotle, of looking for a true definition in order to cognize the essence of the thing.

Aristotle

Since a definition of a thing requires indicating essential features, Aristotle developed an apparatus of language analysis with the purpose of searching for essence. In order to cognize the essence of a thing or what it is, it was necessary to assign it to a certain sort and class. Such a classification procedure was carried out by the analysis of notions. Aristotle is merited with the development of the foundations of logic; and one important contribution to the problem of relation between thinking and existence was his rule: 'existence is not a predicate'. This means that existence cannot be the subject of a logical proof. The method of Aristotle had been used during many centuries by the scholastics up to appearance of Early modern philosophy.

Descartes

Descartes, the founder of the Early modern philosophy, directly connected being with thinking, considering '*cogito ergo sum*' trustworthy. Let someone doubt his existence, but then there is, at least, the doubt. Then Descartes assigns this act of doubt to himself; it follows from this that I am a doubting thing, and hence 'I think therefore I am'. Certainly, the proof of Descartes is erroneous; the impossibility of correct proof follows from Aristotle's result. Let us demonstrate that this proof is neither necessary nor sufficient.

Necessity

Imagine you have the following dream: you see a person in contemplation, who would say to you, "I think therefore I am".

You object, "Dear sir, you do not exist. You are only my dream".

He would respond, "I doubted too that I was, but decided that since I doubted I existed".

You continue to insist: "you with your doubts are only being dreamt by me, but you do not exist, don't you understand?"

He says, "I have thought about it too that I might be mistaken, but since I am mistaken than I am".

An imaginary dialogue can be continued ad infinitum, but since the presence of an act of thinking is accompanied by assigning a subject to the process of thinking, the dispute is unresolvable. With confidence, we can only affirm the presence of a certain thinking act: doubt, mistake and so on.

Sufficiency

The fact that the presence of thinking is not sufficient to confirm one's own existence can be demonstrated by referring to medical practice dealing with the depersonalization-derealization syndrome. The basic forms of depersonalization may concern either the alienation of the outside world, or the alienation of one's own body, or the alienation of one's own self. The patient feels that his sensations and/or actions are detached from him, moved away, are not his own, are lost. At that, the person continues to think and be aware of himself, though in an alienated fashion. Thus, the presence of consciousness is not sufficient for the belief in one's own existence. There are described in the clinical practice such cases when the person consider his thoughts to be not his; the patients may complain that their feelings are separated, are alien to them, are not their own or lost; or experience a sensation that their thoughts or movements belong to someone else or they feel themselves as playing on stage [19]. It is not at all so that the entire reality is experienced by the sufferer as imposed from outside, set up by somebody. No, part of his feelings would be perceived by the person as his own. Nevertheless, the presence of thinking does not guarantee the feeling of existence, since depersonalization is accompanied sometimes by the feeling of derealization, unlikeliness of surroundings, by lost of belief in the existence of one's own self.

While reasoning about thinking and existence, Descartes set a problem, developed not only by philosophy but also psychology and other cognitive sciences. The mind-body problem was one of the basic problems of Early modern philosophy, and to solve it Leibnitz proposed a conception of pre-established harmony.

The *picture theory* is based on implicit usage of the principle of the being and thinking identity and considers experience to be a means of unambiguous translation of the state of affairs into language. To provide such translation, experience must be a simple reflection of the world; to be a source of atomic propositions, experience must be a pure problem-free source of knowledge about the world. Besides, following the tradition of logical atomism, Wittgenstein ascribes the discreteness property to experience, so that each proposition would correspond to one isolated fact only. Such discretization of experience became a source of troubles for both realism and the semantics of color.

Modern Analytic Tradition

In the analytic tradition, a large attention has been paid to the color realism problem, and, hence, to the definition of the notion of color (identity of the being and existence!). The question 'what does it mean to be red (colored)' has been discussed from the position of semantics. Usually, the definitions of red implicitly presuppose a reference to the sensual experience. 'A perceptible object is red iff it looks red in standard circumstances' [22].

Such definition, containing a biconditional term 'red' in both sides, makes us think about vicious circle. The given quotation rather provides us with a definition of standard circumstances, not a definition of red. Because of that, the initial definitions were made more complex; however, a more sophisticated version, whose development was undertaken to avoid the vicious circle, also contained a reference to sensation.

> (RED'3) Y is red' iff Y is a sensation of a phenomenal type Y such that (a), (under standard conditions and in the absence of beliefs to the effect that experiences are not to be trusted) having a sensation of type Y disposes an English-speaking subject to hold 'x is red' true iff x is the cause of y and (b) objects classified as 'red' by English-speaking subjects are disposed to cause sensations of type Y (under standard conditions and in a given subject) [23].

This definition is made more complex by the reference to a phenomenal type, which is an attempt to introduce a division on classes, characteristic for language, into the sphere of the objects of experience. This trick is similar to Russell's theory of types, which divides sets into types in order to avoid the set-theoretical paradoxes. As a whole, such classification is an attempt to fix the way of perceiving the objects in changing conditions; i.e., one more example of ascribing static character to perception, which quite fits the paradigm of the being and thinking identity. From the point of view of passive paradigm, the optimal situation would be the absence of any changes to provide stability in color perception. Such definitions of red should be considered from the point of view of their necessity and sufficiency. To do that, we consider several examples.

1) Necessity

One should "hold 'x is red' true iff x is the cause of y and (b) objects classified as 'red' by English-speaking subjects are disposed to cause sensations of type Y (under standard conditions and in a given subject)". However, nobody would classify as 'red' a signal perceived not by vision but, for example, by the skin surface in standard conditions. Red gives a sensation of warmth, not of red color, if it is sensed by the skin not by eyes. This example illustrates a general statement, known as Müller's doctrine of specific nerve energies. The law of specific nerve energies was first proposed in 1826 by J. Müller who defined perception according to pathway by which the sensory information is carried. The origin of the sensation does not determine the type of sensation. For example, pressing on the eye, action of electric current or chemical irritant all elicit the sensations of flashes of light.

> The same cause, such as electricity, can simultaneously affect all sensory organs, since they are all sensitive to it; and yet, every sensory nerve reacts to it differently; one nerve perceives it as light, another hears its sound, another one smells it; another tastes the electricity, and another one feels it as pain and shock... Sensation is not the conduction of a quality or state of external bodies to consciousness, but the conduction of a quality or state of our nerves to consciousness, excited by an external cause [24].

2) Sufficiency

Let *x* be Mars. In this case, the English-speaking subject would hold '*x* to be red'. However, is it really that Mars is perceived as red? Are there many philosophers that can locate this planet on the starry sky? It looks like an ordinary star, being less bright than Jupiter or Venus. It shines like other stars;

its color is almost as white as of the distant stars. If you look closely and very attentively, you can perceive a slight yellowish tincture. Mars does not look red. From the point of view of color realism, we should not consider the observation conditions to be normal. However, a look to the night sky is a quite ordinary observation, though Mars shines with a color different from the one that is seen on the photo of its surface made by a powerful telescope. Therefore, we cannot conclude that Mars 'causes sensations of type Y' and that 'the object x is classified as 'red' by English-speaking subjects'.

Concluding the historic excursus on the principle of the being and thinking identity, I would emphasize that this principle is in the foundation of the tradition of investigating thinking instead of existence. Thanks to it, philosophers' activity today is devoted to clarifying the notions and, to a great extent, has turned into the debate about words. Philosophers have received many times the indications of the falsity of this principle. For example, Ch. Fillmor (1968) and G. Lakoff (1965) have established, during the development of semantics and syntax, some discrepancies between the structures of consciousness and the outward structures of language.

The history of science abounds in the proofs of this principle's falsity, in particular, the theory of atomic structure of matter, according to which the matter of the atom is concentrated in its center, which contradicts the manifested palpable solidity of the things. Another example is quantum mechanics with its discreteness and imaginary numbers, which so clearly contradicts the common notions about the continuity of processes and their being of real type! The world is arranged differently in comparison to our usual way of perceiving it, whereas our sense organs provide us only with an approximation sufficient for existence in a rather narrow range of conditions.

However, why is it so that philosophy continues to follow the orientation to study language, not the world? The reasons of inertia of the philosophical thinking lie not only in the antiquity of the linguistic tradition but also in the conveniences it provides. First of all, there is postulated the existence of the outside world or objective reality which is considered to be primary and a source of sensations. Secondly, cognizability of the world is presumed through the representational context, reality being the reason of sensation. Both physicalism and representationalism, presuming 'transparency' of perceptual experience, have the 'research aims to identify the material referents of phenomenal concepts' [25]. Finally, the identity of being and thinking ensures the cognition's adequacy.

Thus, the passive paradigm provides an automatic resolution of the basic question of ontology and epistemology. However, the traditional picture theory, together with conveniences, has got some problems which are difficult to resolve within the framework of this conception. Let us mention several of them:

how are the two-dimensional retinal images transformed into three-dimensional bodies;

how is the knowledge possible about something outside of the vision field, for example, the back side of the Moon;

how is the perception of (empty) space possible;

how does a sum of sensations form the bodies;

what is the ontological status of color – a quality which is not an object.

Besides the problems of clarification of some issues, there are critical arguments, many of them addressing too simplified understanding of experience. Slashing criticism of the passive conception had finally led to the situation when the rationalist tradition was left practically without a cognition theory, and experience as a method of obtaining scientific knowledge was discredited.

2.2. Critical Arguments against Experience

The positivists thought that there exists pure experience, expressed by the 'protocol sentences', on whose basis the scientific facts are formulated. Facts are reliable objective evidences, confirming hypotheses. Experience may confirm or refute a hypothesis and, on the basis of experience's data, the decision on falsity or truthfulness of any statement is to be made. However, there have been voiced a number of critical arguments against experience, which undermined the trust to experience as an 'objective independent evidence'. Let us consider in more detail the criticism of the empirical foundations of the scientific method.

Underdetermination of theories by evidence (W. V. Quine).

W. V. Quine (1969) indicated that the sensual data do not determine the scientific theories. It means that one and the same collection of facts may be explained from the point of view of various theories. Since the facts confirm more than one theory, they cannot be brought in as the grounds of preferring one theory to another; that is, cannot be engaged as independent evidences [26].

2) Historical Development of the Scientific Fact (L. Fleck)

In the year of 1935, there was published a book by Ludwig Fleck, in which he considered the history of development of the medical conceptions about deceases. He demonstrated that the perception of data, description of the decease, and its diagnostics depends, in a decisive manner, on the conceptions being dominant in the society. It means that the fact is not invariant and objective evidence, but is a subject of historical development. Correspondingly, truthfulness or falsity of the evidence cannot be absolute but is determined by the context and style of thinking within which it was stated: 'Concerning other proto-ideas, such as the Greek pre-idea of the atom or that of the elements, we are also unable to decide whether they are right or wrong if they are taken out of their chronological context, because they correspond to a different thought collective and a different thought style' [27]. Thus, Fleck points out that fact cannot be independent objective evidence and cannot be considered as a reliable foundation of scientific knowledge.

3) Theoretical Ladenness of Observations (P. Duhem, N. R. Hanson, T. Kun, P. Feyerabend)

The question of ladenness of observations was first raised by P. Duhem in his 1905 book 'The Aim and Structure of Physical Theory'. Describing the physical experiment, he indicated that obtaining the data of the experiment can be subdivided on two parts. The first part is observation; and to observe it suffices to be attentive and have a possibility to perceive. The second part of experiment consists of interpretation of observations which requires the knowledge of many physical theories and the skills of applying them. 'A physical experiment is not simply the observation of a group of facts but also the translation of these facts into a symbolic language with the aid of rules borrowed from physical theories' [28].

Duhem connects theoretical ladenness only with the theoretical interpretation of observations, assuming that the observation itself neither depend on the observer nor requires the knowledge of the theory. A stronger claim was made by Hanson (1958) in his 'Patterns of Discovery'.

Norwood Russell Hanson started the book by considering the following example: two microbiologists are looking at *Amoeba* in a microscope; and, having been asked what they saw, one of them replies that he saw the nuclear, mitochondria, cytoplasm and other cell structure, while the other says that he saw the organs of movement and eating. The discrepancy between the answers is caused by the fact that the first considers amoeba to be a cell, whereas the second, an animal. Is this difference just a consequence of different interpretation of what they saw? Hanson claims that the difference here is much deeper: the microbiologists really see different things.

How is it possible to see different things while observing the same object? The physicist and the Eskimo child, while looking at a roentgen tube, have the same visual data; but the ways of their visual cognition are different, therefore one of them does not perceive the same thing through vision as the other. Observation is not just the presence of a visual experience but also a way, through which this visual experience is acquired. Hanson claims that vision contains more than just sensual data; and the difference of the sensual experience is caused by no means by different interpretations of the seen: vision takes place *before* any interpretation. Theoretical ladenness is a necessary condition of observation, because we tend to see what we expect to see.

The philosophical significance of the thesis of theoretical ladenness is that observation cannot be used for verification of the theory, since it is 'infected' with theoretical assumption and, therefore, there arises the danger of a vicious circle: we compare not the theory with the observations but the theory with the theory. Hence, observation cannot be an objective source of data about the world, since not only the interpretation of observation but the perception itself depends on the theoretical context.

4) Impossibility to Draw a Line between Theory and Experience (G. Maxwell)

The positivists thought that there exists a distinct boundary between the observable and non-observable; moreover, it was considered that the observable was that that can be observed directly, by naked eye. However, in modern science, observation is becoming more and more indirect. When the instruments were less sophisticated, it was considered that they just amplify the sense organs. That which can be viewed in a binocular, can be also viewed by the eyes, if you come close enough. With such understanding, the usage of instruments in observation cannot be a source of problems. However, sophistication of the observation equipment has lead to the situation when many scientific observations can hardly be called 'direct'. There are objects, such as electron, which can be detected only with the help of special devices.

The empiricism dogma about the existence of a distinct boundary between the observable and non-observable was criticized in a Grover Maxwell's (1962) article. The main thesis of Maxwell is that there is no clear-cut boundary between the observable and theoretical. Maxwell begins his argument with a remark that such expressions as 'theoretical entities' and 'dichotomy of the theoretical and observable' are examples of erroneous categories. Terms and conceptions are theoretical, whereas entities can be observable or non-observable. Two questions follow from this distinction: is it possible to split our language into theoretical and non-theoretical parts and can we classify objects on observable and non-observable?

Maxwell gives negative answers to both of these questions. While compiling experimental reports, we use such words as 'mass', 'impulse' and must conclude that our language is laden with theory. Besides, our 'theoretical' knowledge significantly influences our perception. Suppose we show a tennis ball to a Stone Age man. He would not see it being a tennis ball, because to perceive a tennis ball one needs a conception of playing tennis. Thus, our language is substantially laden with theory.

Maxwell also rejects the possibility of subdivision of objects to observable and non-observable, because observability is most connected with the level of the technology development and the power of the instruments used for observation. Therefore, there exists a continuous sequence of the vision processes: vision through air, through a window, through a binocular, through an optic microscope, and through an electronic microscope. Thus, we cannot draw a clear line between the observable and non-observable: while genes were considered to be theoretical objects prior to invention of the electronic microscope, now we can see them in a microscope. Thus, the notion of observability does not presume that the given object can be viewed here and now. The object can be temporarily non-observable for various reasons, but it may become observable under different, more favorable circumstances. Observability is a principal possibility of being seen; thus, x is observable if the conditions exist such that if we put x there we will observe it [29].

To sum up, Maxwell demonstrated that we cannot use the notion of observability as a ground to state that no theoretical objects exist. Maxwell's argument is directed, for the most part, against resolution of the question about the ontological status of objects. Contrary to the positivists who considered real only that is observable, Maxwell insists that we must believe in reality of directly non-observable objects, because observability of an object, postulated by the theory, depends more on the level of technology development than on its nature.

The source of the passive paradigm's reliability – the hypothesis of an isomorphism between the objective reality and its reflection in consciousness – has ultimately led to relativism. Also, though some of the problems of ontology and cognition theory had received natural solutions, there appeared also some problems whose resolution required different conceptions about experience.

2.3. The Myths about Experience: Passivity and Discreteness of Perception

It was necessary for the tradition, which assumed an isomorphism of world and thinking, to formulate such conceptions about experience that one-to-one correspondence would be guaranteed. A passive paradigm, where perception is considered as a direct reflection of the world, was quite convenient for that purpose. Within the framework of the passive paradigm, the main question of philosophy – 'to identify the material referents of phenomenal concepts' – presupposed that a state of mind is caused by our experience (for instance, experience of color), and experience is represented in mind by 'qualia' [30].

a) philosophical myths about experience: passivity of perception

In the end of the XX century, there was appearing more and more often another understanding of experience; the conceptions of passive reflection of reality were substituted by the understanding of experience as active, according to which observation includes motor activity. It was established that perception is impossible without movements; the necessity of motor impulses was experimentally discovered even for hearing – which apparatus is hidden deep inside a bone and, therefore, lacks its own movements.

The paradigm change took place under the influence of criticism; moreover, the main critical arguments put forth against experience were the direct consequences of the perception activity. *Activity of perception* includes searching, gathering, processing and ordering the information coming from the sense organs. Let us show that the above-mentioned criticism is a direct corollary of the perception activity.

The Thesis of Underdeterminacy as a Corollary of Perception Activity

Activity of perception, expressed as *searching* the information relevant for the observer, points to underdeterminacy of the data perception. Since perception is an active search of the information which the subject expects to find and which corresponds to his/her current interests and needs, one and the same situation will be perceived differently by the observer at different moments of time. On a higher level, this selectiveness of search leads to underdeterminacy of the theory by data.

Historical Development of the Scientific Fact as a Corollary of the Perception Activity

Since activity of perception includes *gathering* of the relevant information, which corresponds to observer's expectations, this results in selecting the confirming examples. Each piece of experimental evidence will be interpreted according to the point of view, which the researcher subscribes to. Therefore, the fact possesses cultural relativity; the process of observation and formulation of its result turn out to be dependent on the dominating conceptions. Correspondingly, the change of the world outlook places the accents differently and requires reformulation of the evidence; so the scientific fact is not 'reliable' non-problematic evidence objectively showing the state of affairs in the world. It is a 'subject of historic development'; and both its interpretation and the perception itself depend, in particular, on the views and the world outlook accepted in the society.

Theoretical Ladenness of Observations as a Corollary of the Perception Activity

Sorting the obtained information is performed on the initial stages of perception and includes categorization: first, a rough one (visual image, sound, perception), then a notional one, which according to J. S. Bruner (1974) differs only by the degree of specification, not by nature. Thus, perception always turns out to be notionally laden; and any 'pure' perception is plainly impossible. The categorization process makes perception meaningful, which is confirmed by the examples of 'conceptual blindness' described by the eye surgeons.

A person is describing his first visual experience after the cataract removal:

> He heard a voice coming from in front of him and to one side: he turned to the source of the sound and saw a 'blur'. He realized that this must be a face. Upon careful questioning, he seemed to think that he would not have known that this is a face if he had not previously heard the voice and known that voices came from faces [31].

A man without visual experience 'sees' only the chaos of colored spots, not understanding what is in front of him. This example shows a direct dependence of the real-time perception on the ability of categorization; i.e., the dependence of vision on theoretical interpretation.

Impossibility of Drawing a Boundary between Theory and Experience as a Corollary of the Perception Activity

Processing information includes interpretation and understanding of the perceived. On this stage, important roles are played by the prior experience, intentions, needs and many other factors, which are difficult to control, due to which the descriptions of one and the same situation by different observers are different. Therefore, observation turns out to be so connected with theoretical expectations and axiological orientations that it is impossible to draw a clear line between objective vision and subjective orientations.

b) philosophical myths about experience:
discreteness of perception

The tradition of considering experience as atomic and perception as a single act of receiving a signal of a certain sensor modality takes its origin in the necessity to adjust the conception of experience to the structure of language: a separate word must signify a separate object and a separate statement, a simple isolated fact. Such correspondence makes the understanding of experience compatible with the passive mirror paradigm. The formed tradition to consider experience as a succession of isolated discrete acts provides us with too simplified a picture. Beginning from the logical atomism and up to the first half of the XX-th century, the language was considered as a collection of separate sentences. Experience was also considered as a single isolated fact which can be expressed in the form of protocol or basic sentence.

It should be said in defense of the picture theory that there are physiological premises to treat perception as discrete. In part, such understanding is connected with our possibility of movement, which allows us to constantly change the vision field and constantly perceive something new. The organization of nervous system is such that it reacts more actively on new stimuli, rapidly lowering its activity in habituation. Accordingly, the first impression is the strongest and brightest; and it is this impression that is being referred to while talking about the observation of a fact or a state of affairs expressed by a separate sentence.

Another premise is the ability of vision to single out the objects from the background. A sophisticated complex of sensations coming, for example, from a cup is combined with the traces of the previous experience, as well as with the current goal orientations and thus form an image which we are used to denote by one word – 'cup'. The process of perception of the cup is caused by formation of a complicated sensations complex. The permanent flow of sensations of different kinds is such that no sensation is exactly identical to any other, which is possible due to constant movement of eyes.

> There are three types of eye movements that may be made when tracking visual stimulus... Reflexive saccadic eye movements occur in response to the sudden on set of a peripheral stimulus... Voluntary saccadic eye movements occur under voluntary or planned control... Smooth pursuit eye movements represent a third type of eye movements used in tracking visual stimuli [32].

The eye's fovea changes its position ones in every 150-200 ms. The saccadic eye movements have the capacity to accelerate up to 1000 degrees/second [33]. The saccadic eye movements are very rapid and occur every 100-200 ms. The average saccade takes only 25-30 ms. Taking the day as 12 hours and the duration of every new phase of visual sensation to be 200 ms each, we get that only vision would provide us with more than 216,000 sensations per day. Hearing, skin and osmesis give no less sensations; and there are even more sensations from the muscle movements. And all this numerous psychic acts connect with each other every second and thus form sophisticated complexes of sensations, which we perceive as objects or surfaces, spatial extension or the rays of light.

As it was indicated earlier, it is impossible to explain how the person perceives the bodies while considering experience to be a discrete collection of sensations. However, as the logical atomism afterward gave place to theoretical holism in the form of the Duhem–Quine thesis [34], also for experience the idea of independence of isolated perceptions must give way to a conception taking into account the experience continuity, the mutual influence of sense organs and the role of perceived information in creation of general tuning of the organism. Therefore, I am formulating

The Principle of Empirical Holism

Perception is a continuous in time process of action on the body of a sophisticated complex of irritants of various modality, which exert influence on the entire organism and its motor activity, whose excitation action on the sense organs we experience as sensation due to fixation of our attention on a part of the jointly acting stimuli being a small portion of the complete set of irritants.

On the mind-body language, this principle means that not only the informational component of the perceived signal should be taken into account but the influence of the signals on the condition of the body, since it is the latter that determines our ability to perceive. The body is a necessary mediator between the physical signal and its mental image. The body determines, to a large extent, which ones of the dozens of thousands of sensations acting on the organism each minute are worth of our attention and the action of which should be inhibited. For the philosopher, the body is the dwelling place of the enigma of the mind – brain relations and already because of this it requires attention. For this reason, the next paragraph will be devoted to the influence of the color perception on the body and the discussion of the question how the body can be a source of belief in existence. Let us consider first how the empirical holism principle may be used to overcome the traditional difficulties of the picture theory.

2.4. The Difficulties of the Picture Theory from the Point of View of the Empirical Holism Principle

1. How Does the Sum of Sensations Form Bodies

The jointly acting irritants are responsible for objectivity of our vision; it is precisely thanks to the complex character of perception that we see not the chaos of colored spots but the objects. The mechanism of the sensations connection was clarified in the conditioned reflexes theory of I. P. Pavlov, who indicated that if two different irritations had appeared simultaneously for some time, then a little afterwards the appearance of one of them would induce recollection and reaction to the other. Studying the mechanisms of the conditioned reflexes imprinting, Pavlov noticed that the coinciding in time sensations – visual, acoustic, tactile, gustative, or osmatic – are connected by the brain in such way that the appearance of one of these sensations causes expectation of some other. In such fashion, a photo of a fragrant flower may create a smell sensation. After the sensations of different modalities are combined in an integral complex, the presentation of one of them evokes from memory the idea about the others and there arises an idea about the object. As a result, the constantly concomitant sensations are united into relatively stable entities, so that the action of one of them immediately evokes in memory the entire complex; and so we see the objects, not a bunch of colored spots. The vision on its own is not sufficient for perception of the world; the continuity of experience is necessary combined with the traces of the former experience. Connection of sensations is accompanied by formation of concepts based on the past experience related to manipulation with the given object or action in a given situation. To explain objectivity of vision, we need to consider not an isolated observation but experience in a wider context which includes interaction of the sense organs of various modality and integration with the past experience.

2. How Are Two-Dimensional Retinal Images Transformed into Three-Dimensional Bodies

The skill of vision is formed in early age together with the sense of touch, trying tastes, and manipulations with objects providing the idea of the

> objects extensionality. A prerequisite for the extensionality perception is the sufficient development of the spatial conceptions which are formed, to a large degree, due to observer's activity. The lack of either motional or sensor experience prevents one from perceiving the things as three-dimensional. The effect of seeing the world as flat can be produced artificially; such experiment was described by J. Gibson (1950) who created for himself certain presetting of vision in order to perceive the room not as a room but as a picture. To differentiate between two modes of vision he introduced the notions of visible field and visible world [35].

Perception of objects may change to perception of picture also in the conditions when the normal perception of space is disturbed. Such effect was described by G. M. Stratton (1896) in the experiment with wearing lenses that turn the image upside down and invert it from left to right.

> All images at first appeared to be inverted; the room and all in it seemed upside down. The hands when stretched out from below into the visual field seemed to enter from above. Yet although these images were clear an definite, they did not at first seem to be real things, like the things we see in normal vision, but they seemed to be misplaced, false, or illusory images between the observer and the objects or things themselves. For the memory-images brought over from normal vision still continued to be the standard and criterion of reality. The present perceptions were for some time translated involuntarily into the language of normal vision; the present visual perceptions were used simply as signs to determine how and where the object would appear if it could be seen with restored normal vision. Things were thus seen in one way and thought of in a far different way. This held true also of my body. For the parts of my body were felt to lie where they would have appeared had the instrument been removed; they were seen to be in another position. But the older tactual and visual localization was still the real localization [36].

3. How the Knowledge About the Things Outside of the Vision Field Is Possible, for Example, the Knowledge About the Back Side of the Moon

Since perception has a complex character, the synthesis of information of various modalities gives an integral idea about the objects. Suppose we have never seen the given tomato, but since we have many times before held various other fruits we know that this tomato possesses a back side though we do not see it. Since there is a possibility to turn the visible object or walk around them, the idea of three dimensions becomes an unalienable property of our thinking about the objects.

Various spatial positions of the lightened hemi-sphere of the Moon with respect to the observer on the Earth, which turn the Moon into Crescent, surely indicate three-dimensional character of this body and, extrapolating the experience we have, we may draw a conclusion about this celestial body.

Note that oftentimes quite abstract or even theoretical ideas get involved in the conceptions about the objects. This is indeed so, and the very setting of the question about the other side of the Moon was impossible while the influence of Aristotle's authority was strong and the Moon was viewed as a perfect celestial body fixed on a crystal sphere. Such question could be raised only after the works of Galileo, who established material integrity of the celestial and earthly worlds and this made possible the extrapolation of the experience obtained in the condition of Earth to the celestial bodies. From the positions of empirical holism, to see means to have knowledge of a certain kind; vision is an amalgam of sensations, accumulated experience, expectations and intentions of the observer, theoretical orientations and language.

4. How the Perception of (Empty) Space Is Possible

Space is a complex conceptual-sensor entity, an uttermost generalization of the sensor data of various modalities. The capability of such large-scale synthesis is formed for a person only by the age of 12-13 years provided he/she possesses a sufficiently diverse life experience. For example, the pygmies of Africa, spending most of their life in the jungle, are not able to correctly estimate the distances on the open terrain.

R. Held and A. Hein conducted an experiment with two kittens; one of them was harnessed into a little cart and moved on its own, while the other was sitting tied in the cart. The rest of time the kittens spent in darkness. As a result, the active kitten would see normally; whereas the spatial vision failed to be formed for the passive one. Thus, the sensor-motor experience is necessary to correctly form the space-related conceptions. An important role is played by the exercises in estimation the perspective and changing the distances and the angles of observation of the objects.

At the same time, the sensor apparatus is equipped with special receptors perceiving only the spatial characteristics. For example, the sense of touch gives information about the spatial position of the irritating action; the sense of hearing can localize the sound source in space (binaural hearing); there are the cells in the retina that react on the movement in a certain direction and the cells whose excitation is caused only by presentation of lines of a precise inclination.

Moreover, perception is provided not only by the five, known already from the Antiquity, organs of sense. There is also proprioception and interoception, providing the brain with information about the spatial position of the body, velocities and accelerations experienced by each muscle, and about the body orientation in space with respect to the vertical direction. The movements of the observer give information about the spatial characteristics of the environment; and this information, provided the corresponding experience is present, generates the perception of space. As it was indicated in the beginning, the data of various sensor systems are the foundation of synthesis of the spatial conceptions which development is consummated only in the adolescent age.

The issue of color realism will be discussed in the next chapter.

Summing up this chapter, we should emphasize the necessity of taking into account the integrity of experience, since sensations (even ones we are not aware of) play an important role in the observer's presetting for perception. A practical significance of such presetting is huge; it provides the very possibility of observation and perception. 'In such cases there is usually takes place reorganization of the adaptive reactions, which may create unfavorable conditions for those irritants that are not in the focus of attention at the moment' [37]. The influence of presetting is a general property of perception in general: we are inclined to pay attention only to those details of the surroundings that may serve solving our problems, while we disregard all the rest. However, it is customary now to treat experience as a succession of isolated observations occurring independently from each other. I connect the supposition about the experience discreteness with the philosophical premise concerning an isomorphism of thinking and the world, which I call the principle of the being and thinking identity. To ensure one-to-one correspondence, experience must be simple reflection of the reality. And since thoughts are expressed in the language consisting of separate sentences, experience was ascribed discreteness allowing assigning a separate fact to a separate sentence.

The picture theory was criticized and had some unsolved problems. In this chapter there was formulated the empirical holism principle and, based on it, there were suggested possible solutions of the problems of passive conceptions and the answer to criticism was given.

Chapter 3

3. COLOR REALISM AND OTHER PHILOSOPHICAL ISSUES OF THE MUTUAL INFLUENCE OF SENSE ORGANS

Why is it important for the philosopher to know about the mechanism of perception? Because this knowledge allows shedding new light on the main question of philosophy: the mind-body problem. Since perception is a process during which material signals are transformed into mental images, its investigation allows considering the mind-body relations in a quite localized fashion. The problem is that mind and body have absolutely different properties. In the sphere of the brain the informational processes controlled by electric impulses are dominant. The body is controlled by the electric signals not directly but chemically when there come into the blood the substances which exert stimulating or inhibiting influence on the entire organism. The endocrine glands, which produce neurotransmitters and hormones, are the mediators in passing the information from the brain to the body. The action of the chemical agents is non-local; it covers sufficiently vast domains of the organism. The control of the hormone systems is executed by the subcortical structures of the brain, for example, hypophysis and reticular formation, which influence the general condition of the entire body including the brain.

Signals from the sense organs come first to the subcortical structures, where they are additionally processed. These signals proceed by two paths: specific or nonspecific somatosensory pathways. The first path goes through the thalamus core and serves for the prompt and precise transfer of signals from the sense organs of one modality. The activating influences, coming from the thalamus cores, affect only some limited regions of the cortex. The selective increase of activity of a small region of the cerebral cortex organizes the attention by activating, against the general background, the work of a small number of cortex cells. Usually, the neurons of exactly those sections of the cortex are activated that are engaged at the moment in the most important for the organism activity. This significantly fosters the realization of specific functions of the brain: perception and processing of the most needful signals, and execution of the top-priority motional acts.

The nonspecific afferent paths go through reticular formation, where there takes place the synthesis of information having come from various sense organs. The nonspecific system provides with information the nervous centers responsible for coordination of all the systems of the organism. The ascending influences from reticular formation are slower and cover the entire cortex in a diffusive fashion, causing a general change of its functional condition. The experiments on animals showed that irritation of the nonspecific system increases the general activity level, the level of attentiveness to external signals and improves their perception. The impulses of the nonspecific system play an important role in regulating functional condition of the cerebral cortex neurons. These impulses are especially amplified under the action of new irritations, preparing the organism ahead of time to reacting to an unexpected situation. This always takes place in the investigatory reactions and initial stages of development of a conditioned reflex. The regulation of sensitivity is carried out by the vegetative nervous system which changes its state under the irritants action.

Thus, an external signal never produces an exclusively specific effect manifested in the perception act. Irritation of any sense organ induces at least four responses in the organism: sensor reaction or sensation (under the over-threshold irritation), motor, emotional and vegetative reactions. Correspondingly, color also exerts a complex influence on the organism which allows talking not only about the epistemological role of color but also discussing the ontological issues.

3.1. Color and Realism: Continuity of Experience as a Source of Belief in Existence

Color brings about more realism in the TV film images. The display of colors in the illustrative pictures and on TV screen greatly increases their realism as well as the enjoyment [38]. One of the mechanisms of increasing the color realism is the formation of emotional mood. In such fashion, red color is associated with apprehension; green, with tranquility; blue induces the feeling of coolness and so on. One of biological functions of emotions is amplification of reactions; thus, by inducing emotions, color makes sensations and the experience of the feeling of reality more intense. 'Color is one of most powerful elements in our environment; it is typically the first thing, we notice when we enter a room' [39].

The aim of this paragraph is to discuss the mechanism through which colors increase realism of the perceived. Talking about realism in general, let us quote that existence is not a predicate since the problem of existence cannot be the subject of a logical proof. Existence of a certain object or phenomenon is always a question of belief; and, in the sequel, we will discuss the sources of this belief. Since rational reasoning cannot be a source of belief in existence (and all the proofs of existence contain logical errors), we consider how the experience, being the second source of knowledge about the world, can provide us with such belief.

There are some data that the belief in one's own existence is supported by continuity of sensations. These data come from the neurological practice that deals with the people having no sensation of their own body, no proprioception. The person, who does not feel his own body, seems to himself empty, hulled, and non-existing. There are also the reverse cases when, after amputation, a person continues to feel the amputated body part as his own even months after the operation. Oliver Sacks (1985) describes a case of a sailor with an amputated finger; and it seemed to the patient that this finger was there on its place and it protruded to such an extent that he had to turn his head when he would bring the hands to his face.

The patients with amputated limbs keep experiencing the phantom pains in the lost body part sometimes for months and even years. Therefore, the presence of sensations is not a necessary condition of existence; neither, it must be said, it is a sufficient one; it is just a source of psychological conviction [40].

Taking into account that the continuity of sensations ensures the belief in one's own existence, let us consider in an analogous fashion the issue of belief in the outside world. Let us note that, though each second the receptors of our organism generate hundreds of signals about the presence of sensations, this issue is not simple. First of all, there is important not only the presence of permanent sensations but a certain level of adaptation, because a breakdown or a change of readings in one of the sensor systems may result in the lost of the feeling of reality. For example, Professor George Stratton, who conducted experiments with the lenses turning the images upside down and inverting them, noted the feeling of the visible images being unreal.

> The course of experience was something as follows: all images at first appeared to be inverted; the room and all in it seemed upside down. The hands when stretched out from below into the visual field seemed to enter from above. Yet although these images were clear an definite, they did not at first seem to be real things, like the things we see in normal vision, but they seemed to be misplaced, false, or illusory images between the observer and the objects or things themselves. For the memory-images brought over from normal vision still continued to be the standard and criterion of reality [41].

After a certain period of time, the unreality feeling disappeared; and after several weeks, the visual apparatus adapted to such a degree that the things started to appear normal as they did before the inversion. After the lenses removal, the reverse adaptation proceeded must faster and was not accompanied with the feeling of unreality of the surroundings.

However, experiencing the feeling of unreality of the surrounding world may occur not only due to non-typical irritation of visual apparatus; an analogous effect may be caused by unusual influences to other sensor systems. For example, the Soviet pilots-cosmonauts, describing the action of weightlessness on the organism, note that a certain percentage of people are losing the feeling of events reality when they come to the weightlessness conditions. Some people are experiencing complete disorientation and self-identity loss. For example, one of the tested persons, observed by the doctor, experienced a motor excitation, cried and his face had an expression of horror. Afterwards, he described his condition as follows: 'I did not realize that the conditions of weightlessness had come. Suddenly, I had a sensation of rapid downward falling; and it seemed to me that everything around was breaking down, decomposing and flying to all sides. I was taken by a feeling of horror; and I did not understand what was happening around me' [42]. This example, showing the reaction of the organism to the lack of habitual irritation of vestibular apparatus, also illustrates the presence of mutual influences of sense organs.

Now, let us turn to the question of how does color contribute more realism to images. Certainly, colors are an extra source of sensations and already by that they increase the reality feeling. However, this is not their only action; there are some other ways through which colors influence the organism.

For example, Faber Birren in his book 'Selling with Color' mentions on the effect that colored lights had on blindfolded subjects. In this instance the subjects stood before the light with arms extended straight out in front. Under the influence of red light, the subjects' arms spread apart; under the influence of green light, the arms tended to come together. This is an instance where color has worked independently of vision [43]. Hence, the electromagnetic radiation of various wavelengths, whose reaction with the retina is perceived by us as the sensation of color, exerts a complex influence on the organism. As it was already said in the first part, color acts on the vegetative nervous system, accelerating or inhibiting the biochemical processes and so exerts either exciting or inhibiting influence.

> It has been noted that warm colors are more stimulating than cool colors. Muscular activity was measured under various lights and showed the following effects. Under ordinary light muscular activity registered 23 units, under blue light 24 units, under green 28, under yellow 30, under orange 35, and under red 42 units. Tests indicated that students made greater progress in arithmetic while working under a brilliant red light than they made when under ordinary illuminatia [44].

Correspondingly, while discussing the functions of color vision, we must take into account the complex action of the signal on the organism, its role in the formation of emotions. Emotion is a very old acquisition of the organisms in the process of evolution. It is an integrated reaction which allows evaluating promptly the quality of the irritant: whether it is beneficial or harmful, whether one has to seek it or avoid it. An emotional reaction arises well in advance of the complete recognition of the image and the formation of the actions program. The biological role of emotions is that they are a mechanism of amplification of the reactions. Thanks to emotion, a slight negative irritation may induce a powerful response. Color, as one of the factors of influence on the vegetative nervous system, has a direct access to the emotional sphere and, thus, it amplifies emotions and increases the feeling of realism.

It must be recognized that brightness and color strongly affect the attitudes and reactions of the entire human organism. High brightness and strong color stimulation tend to excite the body to increase muscular tension, blood pressure, and respiration. On the other hand, dimness and suppression of color tend to relieve tension and to relax the autonomous nervous system. Brightness, therefore, is conducive to muscular activity, while softness of color is conducive to mental activity [45].

Summing up, let us indicate the main mechanisms of color influence on the increase of the realism of the perceived. First of all, colors are an additional source of perceptions, the continuity of which ensures the feeling of reality. Color exerts an additional influence on the nervous system, either increasing or inhibiting its activity, even without participation of vision. A strong connection between color and the mechanism of emotions provides amplification of the organism reactions to color, which additionally influences the nervous system and increase the reality feeling. Besides, color participates in the formation of presetting of the organism to the perception of other signals. Let us consider further the cognitive functions of color.

3.2. The Color and Cognition

Our ability to distinguish colors turns out to be one of the cognitive capabilities, which makes our orientation in the world easier. From the point of view of empirical holism, we must consider not only the process of color perception but also the influence of the color action on the emotional sphere, motor and vegetative systems. The fist two ways of the color influence on the cognitive processes have been widely investigated and are known. 'Color adds dimension to a world that visually would otherwise consist only of light and dark, form and motion. With color, we can better tell the time of day, the ripeness of fruit, the identities of various plants and animals, and our state of health' [46]. Colors help us distinguish the things out of the background and, thus, accelerate the orientation in the surroundings. For example, 'conjunctions of color and depth in a visual search task is faster than the detection of targets defined by a single attribute' [47].

The importance of color vision in the 'primitive' society of hunters and gatherers was so high that there were much less people suffering from colorblindness among those tribes, because they were subjected to the natural selection. Still, the frequency of colorblindness is smaller in the less civilized countries. One in twelve Caucasian (8%), one in 20 Asian (5%), and one in 25 African (4%) males are so-called 'red-green' colorblind [48].

In the modern society, the color vision plays a smaller role, and the percentages of the cases of 'red-green' colorblindness are approximately 8% among males and 0,5% among females, while three-fourth of the people with the defects of color vision have a diminished sensitivity only to one color. However, color is not a necessary means of the world cognition, though provides some extra advantages. That the capability of distinguishing colors is not a very significant cognitive means is witnessed by the fact that the people with the color vision defects sometimes do not suspect them until the anomalies are discovered with the help of simple laboratory tests. Since the person cannot know which colors are perceived by other people, the manifest confusion of colors is not so easy to reveal during an oral enquiry. A person with a defect of vision who confuses red and green colors can recognize the traffic lights by their brightness and position. Experience has taught him to call a darker color red and a lighter one, green; even if he does not see the real difference in hues [49].

Less is known about the influence of color on the cognitive abilities through its action on the vegetative and motor systems; and we consider this issue in more detail. From the point of view of enactive approach, the cognitive function of color, realized through its action on the motor system, is that color forms expectations and reinforce them. In this context, expectation means readiness to perception which includes a necessary presetting of the sensor apparatus. The abilities of perception and movement had undergone a lengthy co-evolution; they intertwine so closely that movement without perception and perception without movements are both impossible. The very process of perception proceeds not as passive imprinting but as an active from beginning to end process [50]. On the basis of the planned program of actions, the central nervous system carries out presetting the excitability of all the engaged sensor and motor elements [51]. The anticipatory actions are possible on the basis of foresight, a plan of actions thanks to which there takes place extrapolation and correction; the anticipatory movements are connected with changing the neuromuscular tone. The muscular presetting plays a significant role in the acceleration of recognition of what is seen; and color plays an important role in creating such presetting. Color is a characteristic property of objects and fosters their rapid recognition; for example, the colors of the dandelion, lemon and chick are different; and the perception of the corresponding tint of yellow creates readiness to perceive the corresponding object. Thanks to the color vision, we orientate better in space, because blue hues seem to come from farther things and warm hues, from closer ones. 'Blue is driven forward by the white ground, red-orange also, and yellow stands out only slightly from the white' [52]. The spatial conceptions are extremely important for organization of future movements aimed at solution of sensor or motor tasks.

Perception is subordinated to the purpose-oriented behavior of the subject: those impulses we do not care about are simply not being noticed. The goal orientation determines the criteria of selection of information, relevant for the problem solution or the goal achievement. In order to demonstrate the importance of prior setting for perception, we consider an example.

Example of Presetting Influence on the Possibility of Observation

Newton held to the hypothesis of the corpuscular nature of light, that is, light being a bunch of linearly moving particles. Prior to Newton's works, the phenomena of interference and diffraction was investigated and described in detail by F.M. Grimaldi, R. Hooke, and Ch. Huygens. However, 'Newton seems to have ignored or overlooked diffraction effects of the use of a small hole as image, though these had been noticed by his contemporaries' [53]. At the same time, he had a possibility to observe the manifestation of the light ray diffraction, when there take place deviations of the ray from linear path and enlargement of the shadow thrown by a small object. Nevertheless, Newton had not noticed the inner strips of light appearing due to diffraction in the object's shadow. Augustin Fresnel wrote that it is difficult to understand how the light bending into the inner part of the shadow could have eluded such an experienced observer as Newton. Most probably, it happened because of Newton's theoretical convictions, which, to a certain extant, closed his eyes on the significant events that made the standing of his principle weaker.

> The experiment of Grimaldi on the crested fringes within the shadow, together with several others of his observations equally important, has been left unnoticed by Newton. Those who are attached to the Newtonian theory of light, or to the hypothesis of modern opticians founded on views still less enlarged, would do well to endeavor to imagine anything like an explanation of these experiments derived from their own doctrines; and if they fail in the attempt, to refrain at least from idle declamation against a system which is founded on the accuracy of its application to all these facts, and to a thousand others of a similar nature 54].

This example shows the importance of preliminary setting for the perception process. Changing the physical readiness to perception may create unfavorable conditions for those irritants which are not in the center of attention. When the readiness to perceive a phenomenon is absent, it may remain unnoticed; and this is especially true for low-observable effects.

Besides, color exerts an additional activating influence on the nervous system. Since all excitations from the sense organs go through the center that regulates the tuning of vegetative nervous system, the color influence may cause changes involving the entire organism. At that, color, as a source of additional irritation of the sensor system, oftentimes exerts an activating influence; and this favorably affects the perception of signals of other modalities. According to the rule of concentration of additional activation in proportion to the existing excitation of nervous centers, there takes place stronger concentration of attention on the background of general activity. Since every second the organism is being bombarded by a huge number of external irritants the majority of which miss the consciousness, an additional influence of color may increase the effectiveness of perception in general. As an example of positive influence of color on increasing the perception capability, we may point to a method applied during the World War II when, before the night flights, the airplane pilots would spend several minutes in a room lighted by red color. Such training exerted an activating influence on twilight vision.

Summing up, let us indicate that the ways of the color influence the organism are manifold and are not limited to seeing the object as colored. The complex character of the color influence is manifested also in organization of general activity of the central nervous system, influence on muscular tone, formation of the readiness to perception, and emotional coloring of the events. In my opinion, the answer to the question 'what does it mean to be colored?', presupposing the discussion of ontological and epistemological aspects, must take into account the integral and complex character of perception.

CONCLUSION

For the color realism tradition, the constant change of the conditions of color perception and the adaptive changes of sensitivity connected with it pose a problem. From the standpoint of the linguistic tradition, it would be convenient if we perceived red always the same way. The question concerning the changes of color sensitivity was a source of predicaments and is not considered. It is being disguised in the 'normal conditions' expression. The understanding of normal conditions is rather wide and, due to its uncertainty, allows partly hiding the difficulty of solving the problem of color's ontological status. The light of a certain wavelength under its action on the cone-cells apparatus of the eye gives rise to the sensation of color; thus, the color experience is physically caused. However, the knowledge of the physical nature of light does not give understanding of what is the vision of red; there is an explanatory gap here.

We have rejected the traditional way and discussed the mechanisms of changing the color vision sensitivity and the mutual influences of various sense organs. Understanding of perception as a continuous process and taking into account the complex character of the color action on the organism allows shedding light differently on some ontological and theoretical-cognitive questions. Under the action of external irritants on the organism, not only some mental states arise but there are taking place some changes of the condition and motor activity of the entire body which points out to closer connection between mind and body than the modern philosophy assumes. The body transforms physical signals, coming from the outside world, into mental images or a certain state of mind; and the mechanisms of such transformation require the attention of philosophy.

In combination with enactive approach, the empirical holism principle was applied to the solution of some problems and the response to the critique of empiricism. Some approaches were outlined to solution of ontological problems from the standpoint of experience's continuity. The applied approach requires further development both in breadth, for example, to explain the phenomenon of speech and symbolic function of language, and in depth, to clarify better the process of interaction of mind and brain. Clearly, the precipice, which the person feels as the difference between the object and its mental image, should be considered in the context of the cause-and-effect connections. Brain and body have different languages. Nervous system uses electric signals to pass the messages. Body speaks the language of chemistry; even such electromagnetic by its nature process as the action of light ray to the retina, undergoes, on its way to cerebral cortex, a chemical transformation causing decomposition of the light-sensitive compounds like rhodopsin. Chemical processes are closely connected with electrical ones, being different from the latter by the non-local character of action. Therefore, any signal, for example, color, becomes the source of not only a mental image but has a wide opportunities to influence the general condition of the organism, including changing the very possibility of perception.

The analytic tradition, which concentrates its efforts on studying thinking and language, is unable to take into account such influences and they turn out to be excluded from the vision field of philosophers. However, studying the perception mechanisms may offer some interesting perspectives for understanding of the mind and body relation. Color is a promising theme within the framework of studying cognitive processes with the purpose of building bridges across the explanatory gap.

Notes

[1] Newton 1979 p. 4.

[2] Harré 1981.

[3] Young 1859 pp. 51-52.

[4] Schwartz 1946 vol. 3.

[5] Piéron 1945.

[6] Kravkov 1950 p. 345.

[7] Lebedinskii, Pressman, Fadeeva 1948 pp. 104-111.

[8] Mkrtycheva L. I., Samsonova V. G. 1948.

[9] Kravkov 1948 p. 17.

[10] Ibid p. 110.

[11] Katz, Schmidt 1940.

[12] My translation from Russian: Kravkov 1948 p. 110. In original: "в большей степени притекает туда, где уже имеется значительное возбуждение, и в меньшей мере добавляется к слабо возбужденным участкам нашего зрительного аппарата".

[13] Ukhtomskii 1925 p. 60.

[14] My translation from Russian: Kravkov 1948 p. 88. In original: "В подобных случаях обычно происходит перестройка приспособительных реакций, могущая создать неблагоприятные условия для тех из раздражителей, которые не находятся в данный момент в фокусе внимания".

[15] My translation from Russian: Kravkov 1948 p. 110. In original: "наши ощущения могут… заметным образом меняться в зависимости от действия на субъекта побочных раздражителей, настолько слабых самих по себе, что сами они каких-либо ощущений не вызывают".

[16] Curd 2004 p. 58-59.

[17] Ibid. pp. 34-36.

[18] Dancy 2004 p. 36.

[19] The description of the symptom and the examples see: Simeon Abugel 2006; Sadock 2008 p. 296.

[20] Leibniz 2008 p. 196.

[21] Hacker 2001 p. 147.

[22] Peacocke 1984 p. 365.

[23] Glüer 2007 p.124.

[24] Müller 1844. s. 667. Translated by Edwin Clarke and Charles Donald O'Malley. The original text: „Sind die Nerven bloss Leiter für die Eindrücke des Lichtes, der Tonschwingung, der Riechstoffe wie kommt es, dass derjenige Nerve, welcher die Riechstoffe riecht nur für diese Art von Eindrucken ist, für andere nicht, und dass ein anderer Nerve die Riechstoffe nicht riechen kann; dass dor Nerve, welcher die Lichtmaterie oder die Oscillationen derselben empfindet, die Oscillationen der schallleitenden Körper nicht empfindet, der Gehörnerve für das Licht, der Geschmacknerve für die unempfindlich ist, der Gefühlsnerve die Schwingungen Körper nicht als Ton, sondern als Gefühl von Erbitterungen. Diese Betrachtungen haben die Physiologen genöthigt, einzelnen Sinnesnerven eine specifische Empfänglichkeit für Eindrücke zuzuschreiben, vermöge welcher sie nur Leiter gewisse Qualitäten, nicht aber für andere sein sollten".

[25] Papineau 2003 p. 207.

[26] Quine 1969 pp. 26-68.

[27] Fleck 1981 p. 25.

[28] Duhem 1954 p. 156.

[29] Maxwell 1962 p.16.

[30] Shoemaker 2003 p. 253-255.

[31] Gregory Jean 1963 p. 366.

[32] Nelson 2001 p.329.

[33] Erickson 2007 p. 58.

[34] Duhem (1954) wrote: "An experiment in physics can never condemn an isolated hypothesis but only a whole theoretical group". p. 183.

[35] Gibson 1950 p. 16.

[36] Stratton 1896 pp. 613-614.

[37] My translation from Russian: Kravkov 1948 p. 88. The original text is: "В подобных случаях обычно происходит перестройка приспособительных реакций, могущая создать неблагоприятные

условия для тех из раздражителей, которые не находятся в данный момент в фокусе внимания".

[38] Padham Sonders 1978 p. 47.

[39] Venolia 1988 p. 57.

[40] Mitchell 1872 p. 318.

[41] Stratton 1896 pp. 613-614.

[42] My translation from Russian in: Leonov, Lebedev 1966 p. 7. The original text is: "Я не понял, что наступило состояние невесомости. У меня внезапно возникло ощущение стремительного падения вниз, и мне показалось, что все вокруг рушится, разваливается и разлетается в стороны. Меня охватило чувство ужаса, и я не понимал, что вокруг меня происходит".

[43] Abbot 1947 p. 131.

[44] Ibid p. 129.

[45] Birren 1959 p. 94.

[46] Venolia 1988 p. 57.

[47] Christophe Guibal Dresp 2004 p. 32.

[48] Okabe http://jfly.iam.u-tokyo.ac.jp/color/

[49] Padham Sonders 1978. pp. 212-213.

[50] Bernstein 2004 p. 486-487.

[51] Ibid p. 473.

[52] Itten 1970 p. 78.

[53] Harre 1981 p. 184.

[54] Fresnel 1859 p. 73-74.

REFERENCES

Abbot, A. G. (1947). *The color of life*. New York: McGraw-Hill.

Birren, F. (1959). Functional use of color. In: *Field Applied Paints and Coatings.* National Research Council (U.S.), National Research Council (U.S.). Building Research Institute, 93-96.

Bruner, J. S. (1974). *Beyond the Information Given: Studies in the Psychology of Knowing* Ed. J.M. Anglin London: Allen and Unwin.

Carol, V. (1988). *Healing Environments: Your guide to indoor well being.* Berkley: Celestial Arts.

Christophe, R. C., Guibal, J. & Dresp, G. B. (2004). Interaction of color and geometric cues in depth perception: When does ''red'' mean ''near''? *Psychological Research, 69*, 30–40.

Curd, P. (1998). *Legacy of Parmenides: Eleatic Monism and Later Presocratic Thought.* Parmenides Publishing Princeton University Press.

Dancy, R.M. (2004). *Plato's Introduction of Forms.* Cambridge: Cambridge University Press.

Dobryakova, O. A. (1944). Influence of gustative, temperature and auditory irritants on the critical frequency of blinking of mono-chromatic rays. *Problems of physiological optics*, 1944, vol. 2. Добрякова, О. А. (1944). Влияние вкусовых, температурных и слуховых раздражителей на критическую частоту мелькания монохроматических лучей. *Проблемы физиологической оптики, 2,* 42-53.

Dobryakova, O. A. (1948). On some possibilities of conditional-reflex changing the sensitivity of vision. *Problems of physiological optics*, *6*, 308-313. Добрякова, О. А. (1948). О некоторых возможностях условно-рефлекторного изменения чувствительности зрения. *Проблемы физиологической оптики, 6,* 308-313.

Duhem, P. (1954). *The Aim and Structure of Physical Theory.* Princeton: Princeton University Press. (translated from the second edition published in 1914 by P. Wiener).

Erickson, G. B. (2007). *Sports Vision: Vision Care for the Enhancement of Sports Performance*. St. Louis: Butterworth Heinemann Elsevier.

Fillmore, Ch. J.: (1968). The Case for Case. In E. Bach, R. T. Harms (eds.), *Universals in Linguistic Theory* (1-88). New York: Holt, Rinehart and Winston.

Fleck, L. (1981). *Genesis and Development of a Scientific Fact.* Chicago: Univ. of Chicago Press. (F. Bradley, Trans.).

Fresnel, A. (1859). Memoir on the Diffraction of Light. In: H. Crew (ed.) *The Wave Theory of Light Memoirs by Huygens, Young and Fresnel.* (79-144) New York: Cincinnati: Chicago American Book Company.

Gibson, J. (1950). *The perception of visual world.* London - NY: Prentice-Hall.

Glüer, K. (2007). Colors without circles? *Erkenntnis*, *66*, 107–131.

Gregory, R. L., & Jean G. (1963). *Wallace Recovery from Early Blindness: A Case Study.* Cambridge: W. Heffer and Sons.

Hacker P.M.S. (2001). *Wittgenstein: connections and controversies*. Oxford: Clarendon Press.

Harré, R. (1981). *Great scientific experiments: Twenty experiments that changed our view of the world.* Oxford: Phaidon Press Ltd.

Hartridge, H. (1947). The visual perception of fine detail. *Philosophical Transactions of the Royal Society of London*, 232, 519–671.

Held, R., & Hein, A. (1963). Movement produced stimulation in the development of visually guided behavior. *Journal of Conparative and Physiological Psychology*, 56, 873-876.

Itten, J. (1970). *The Elements of Color: A Treatise on the Color System of Johannes Itten Based on His Book "The Art of Color".* B. Faber (ed.), Van H. Ernst (trans.). New York, Van Nostrand Reinhold.

Katz, B. & Schmitt, O. H. (1940). Electric interaction between two adjacent nerve fibres. *J. Physiol, 97 (4),* 471–488.

Kravkov, S. W. (1934). Changes of visual acuity in one eye under the influence of illumination of the other or of acoustic stimuli. *J. &per. Psychol.,* 17, 805-812.

Kravkov, S. W., & Semenovskaja, E. N., (1933). Steigerung der Lichtempfindlichkeit des Auges dutch vorangehende Lichtreize, *Arch. Ophth.*, Leipsic.

Kravkov, S.W. (1941). Colour vision and the autonomic nervous system. *J. Opt. Soc. Amer., 31*, 335-337.

Lakoff, G. (1965). *On the nature of syntactic irregularity.* Cambridge, Mass.: Harvard University, Computation Laboratory.

Leibniz, G. W. 1898 (2008). *The monadology and other philosophical writings.* Translated with introduction and notes by Robert Latta. Oxford: Oxford University Press.

Maxwell, G. (1962). The Ontological status of theoretical entities. In H. Feigl and G. Maxwell (eds.) *Minnesota Studies in the Philosophy of Science*, Vol. III: Scientific Explanation, Space, & Time. (3-27). Minneapolis: University of Minnesota Press.

Mitchell, S. W. (1872). *Injuries of nerves and their consequences.* Philadelphia: J. B. Lippincott.

Mkrtycheva, L. I. & Samsonova, V. G. (1948). Change of the color sensation under the hypoxemia influence. *Problems of physiological optics, 3*, 120-129. Мкртычева, Л. И. & Самсонова, В. Г. (1948). Изменение цветоощущения при воздействии гипоксемии. *Проблемы физиологической оптики, 3*, 120-129.

Müller, J. (1840). *Handbuch der Physiologie des Menschen für Vorlesungen.* J. Hölscher (ed.). Trans. E. Clarke and C. D. O'Malley. http://www.experiencefestival.com/a/Johannes_Peter_Mller_-_Quote/id/5188063.

Nanson, N. R. (1958). *Patterns of Discovery: An inquiry into the conceptual foundation of science.* Cambridge: Cambridge University Press.

Nelson, C. A. (2001). *Handbook of Developmental Cognitive Neuroscience.* Cambridge: MIT Press.

Newton, I. (1979). *Opticks: Or, A Treatise of the Reflections, Refractions, Inflections & Colour of Light.* New York: Courier Dover Publications.

Noë A. (2004). *Action in Perception.* Cambridge, MA: The MIT Press.

Okabe, M. (2002). (modified on 2.15.2008, 9.24.2008). Color universal design (CUD) How to make figures and presentations that are friendly to Colorblind people –Jikei Medial School (Japan) http://jfly.iam.u-tokyo.ac.jp/color/

Padgham, C.A. & Saunders, J. E. (1975). *The Perception of light and colour.* London, G. Bell and Sons.

Papineau, D. (2003). Could there be a science of consciousness? *Philosophical Issues, 13*,205-220.

Peacocke, C. (1984). Colour concepts and colour experience. *Synthese, 58*, 365-381.

Piéron H. (1945). *La sensation, guide de vie.* Paris: Gallimard. Piéron H. (1952) The sensations: Their functions, processes, and mechanisms (M. H. Pirenne & B. C. Abbott, Trans.). New Haven, CT: Yale University Press.

Quine W. V. O. (1969). *Ontological Relativity and Other Essays*. New York: Columbia University Press.

Sacks, O. (1985). *The man who mistook his wife for a hat.* London: Duckworth.

Sadock, B. J (2008). *Kaplan and sadock`s concise textbook of clinical psychiatry*. Manufacturer: Lippincott.

Schwartz, L. A. (1946). On the phenomena of the color vision sensitization. *Problems of physiological optics*, *3,* 33—39. Шварц, Л. А. (1946). О явлениях сенсибилизации цветного зрения. *Проблемы физиологической оптики, 3,* 33—39.

Schwartz, L. A. (1948). Change of the color sensation in the emotional states of vision. *Problems of physiological optics*, *6*, 314-318. Шварц, Л. А. (1948). Изменение цветоощущения в эмоциональных состояниях. *Проблемы физиологической оптики, 6*, 314-318.

Schwartz, L. A. (1949). Influence of the colored illumination on the hearing sensitivity under various condition of the person. *Problems of physiological optics, 7,* 53-56. Шварц, Л. А. (1949). Влияние цветного освещения на чувствительность слуха при различных состояниях человека. *Проблемы физиологической оптики*, *7,* 53-56.

Okabe, M. (2002). (modified on 2.15.2008, 9.24.2008). Color universal design (CUD) How to make figures and presentations that are friendly to Colorblind people –Jikei Medial School (Japan) http://jfly.iam.u-tokyo.ac.jp/color/

Shoemaker, S. (2003). Content, character and color. *Philosophical Issues, 13(1),* 253-278.

Simeon, D. & Abugel, J. (2006). Feeling Unreal*: Depersonalization Disorder and the Loss of the Self.* NY: Oxford University Press.

Stratton, G. M. (1896) Some preliminary experiments on vision without inversion of the retinal image. *Psychological Review*, *3(6),* 611-617.

Ukhtomskii, A. A. (1925). The dominant principle. *New in reflexology and physiology of nervous system* (60—66). Leningrad-Moscow. Ухтомский, А. А. (1925). Принцип доминанты. *Новое в рефлексологии и физиологии нервной системы.* (60—66). Л.- М.: Госиздат.

Young, T. (1859). On the theory of light and colors. In: H. Crew (Ed.) *The wave theory of light: memoirs by Huygens, Young And Fresnel.* (45-61) New York: Cincinnati: Chicago: American Book Company.

Zaretskaya Z. (1941). B. The action of the color secondary irritants on the intraocular pressure. *Problems of physiological optics*, *1*, 35-37. Зарецкая, З. Б. (1941). Действие цветных раздражителей на внутриглазное давление нормального и патологического глаза. *Проблемы физиологической оптики,.1*, 35-37.

Кравков, С. В. (1948). *Взаимодействие органов чувств.* М. -Л.: Изд-во АН СССР.

Кравков, С. В. (1950). *Глаз и его работа.* М. -Л.: Изд-во АН СССР.

Кравков, С. В. (1951). *Цветовое зрение* М.: Изд-во АН СССР.

Лебединский, А. В., Прессман, Я. М. & Фадеева, А.А. (1948). Новые данные о взаимодействии центра и периферии сетчатки. *Проблемы физиологической оптики, 6*, 104-111.

Леонов А.А.& Лебедев В. И. (1966). Об отражении человеком пространства вне Земли. *Вопросы философии, 1*, 3-11.

Орбели, Л.А. (1935) *Лекции по физиологии нервной системы.* Ленинград: Биомедгиз.

Шеварева, В. К. (1950). Влияние цветного освещения на мускульно-двигательную работоспособность. *Проблемы физиологической оптики, 9,* 127—130.

INDEX

D

E

F

G

H

I

J

L

M

N

O

P

Q

R

S